Sustainable low energy cooling: an overview

CIBSE Knowledge Series: KS3

Principal author
Chris Parsloe

Editors
Gay Lawrence Race
Ken Butcher

CIBSE

Contents

Note from the publisher

This publication is primarily intended to provide guidance to those responsible for the design, installation, commissioning, operation and maintenance of building services. It is not intended to be exhaustive or definitive and it will be necessary for users of the guidance given to exercise their own professional judgement when deciding whether to abide by or depart from it.

▌1   Introduction

This publication gives an overview of the main low energy cooling options available to building services engineers. Its aim is to promote awareness of the alternatives available and their suitability for different applications.

It is aimed at a target audience of relevant non-experts such as clients, architects and students. However it should also be of use to more experienced engineers as an accessible overview which can be used in discussions with clients and other members of the design team.

1.1   The need for low energy cooling

There is compelling evidence that the UK climate is changing due to global warming, and that average UK temperatures can be expected to increase by 4–6 °C over the next 50–80 years. This will result in a higher frequency of summertime temperatures in the range 30–35 °C. It is also predicted that many existing buildings could suffer from occasional over-heating by the 2020s.

In order to help delay the effects of global warming, it is likely that increasingly punitive measures will be introduced against high energy consumers. The Climate Change Levy, introduced in April 2001, is the first indication of this.

Low energy cooling strategies are therefore essential in order to:

— minimise mechanical cooling loads and hence reduce green house gas emissions and energy costs for new buildings

— avoid the risk of future over-heating in new and existing buildings in the event of warmer summer temperatures.

1.2   Strategies for achieving low energy cooling

There are a variety of ways in which the mechanical cooling load in buildings can be reduced. Many of the measures available require little additional expenditure. The main strategies for reducing cooling loads, in order of preference are:

— reduce unnecessary heat gains to buildings

— adopt passive cooling solutions, i.e. solutions which utilise various features of the building to generate and store cooling energy

— utilise cooling energy from naturally occurring renewable sources local to the building

— install mechanical cooling plant but utilise all available opportunities for free cooling during operating periods when full mechanical cooling is not required

The solutions described in this publication are categorised within each of these four strategy headings under in sections 2, 3, 4 and 5.

In addition, a further section is included (section 6) which describes some of the cooling delivery systems that are best able to maximise the energy saving potential of the cooling solutions described in preceding sections.

Table 1 gives an overview of the cooling solutions described in this guide, and provides an indication of their relative merits.

Table 1:

Comparative merits of different low cooling solutions

Low cooling solution	Energy saving potential	Cost to implement	Ability to provide low chilled water temps.*	Ability to provide accurate control of space temps.	Design/ operating risk
Reduce heat gains	☺	☺	NA	NA	☺
Natural ventilation	☺	☻	NA	☹	☻
Mixed mode ventilation	☺	☻	NA	☹	☻
Night cooling	☺	☻	NA	☻	☻
Ground cooling air system	☺	☹	NA	☻	☻
Ground cooling water system	☺	☹	☻	☻	☻
Surface water cooling	☺	☻	☻	☻	☻
Free cooling in re-circulating air systems	☻	☺	NA	☺	☺
Free cooling in full fresh air systems	☻	☺	NA	☺	☺
Desiccant cooling	☻	☻	NA	☺	☺
Free cooling in systems with cooling towers or evaporative coolers	☻	☺	☻	☺	☺
Thermosyphon chillers	☻	☺	☺	☺	☺

☺ = good   ☻ = average   ☹ = poor   NA = not applicable

* Inability to achieve low chilled water temperatures indicates that the solution is best suited for use with a delivery system which can utilise raised chilled water temperatures, such as those described in section 6.

2   Design to reduce building heat gains

The simplest way to reduce the requirement for mechanical cooling systems is to design the building envelope such that heat gains to internal spaces are minimised. Many of the options for reducing cooling load can be achieved with minimal impact on construction costs.

High performance building envelopes are described in CIBSE TM29: *HVAC strategies for well-insulated airtight buildings*[2]. TM29 explains how relatively small improvements in thermal insulation, glazing and air infiltration (beyond the requirements of the Building Regulations Part L (2002)) can result in greatly reduced heat gains and consequent cooling loads.

Minimising a building's cooling requirement can result in a significant additional benefit in that many of the low energy cooling strategies described in subsequent sections of this guide become more viable. In some cases refrigeration plant can be eliminated altogether.

The main features of high performance building envelopes, as described in TM29, are summarised in Table 2.

Table 2:

Achieving high performance building envelopes

Feature	Achievable by:
Better thermal insulation: average wall U-values of not greater than 0.3 W/m²·K	• Attention to building design and construction — emphasis should be placed on the importance of: — eliminating cold bridging between inner and outer wall surfaces, i.e. insulation layers must be continuous and without conductive materials passing through them — inspecting the quality and continuity of insulation before it is covered over and inaccessible — thermal imaging of the completed building to demonstrate the effectiveness of thermal insulation or identify areas requiring remedial measures.
Improved window design: window U-values of 1.0–2.0 W/m²·K	• Selection of windows with carefully designed frame and glazing arrangements. Low U-value windows are available which utilise: — increased air gaps between panes — argon-filled gaps between panes — low emissivity glass — the addition of a further thermally isolated pane (e.g. a '2+1' type window which incorporates a single outer pane thermally isolated from an inner double glazed unit).
Solar shading: window shading coefficients of 0.2–0.3	• Incorporation of glazing with: — internal blinds — mid-pane blinds between the panes of a double glazed unit — ventilated cavity blinds, e.g. between the panes of a '2+1' where the cavity between the outer pane and the double glazed unit is ventilated so that solar radiation absorbed by the blind is vented to outside — external blinds — surface coatings on the glass.
Reduced air leakage: air permeability levels of less than 5 (m³/h)/m² at 50 Pa pressure differential	• Attention to building design and construction to minimise air leakage — emphasis should be placed on the importance of: — incorporating draught lobbies at external door openings — incorporating continuous air barrier elements within the structure with due attention as to how air seals will be maintained — training for site operatives regarding the aims and importance of air barriers — inspection for air barrier quality and continuity before it is covered over and inaccessible — building pressure testing to demonstrate the effectiveness of the air barrier insulation or identify areas requiring remedial measures.

▌3   Passive cooling solutions

Passive cooling solutions utilise various features of the building itself, and its surrounding environment to generate and store cooling energy. The most common approaches to passive design are the use of flexible, intelligent ventilation strategies. These strategies can be further enhanced by the use of night storage using the thermal mass of the building itself to store cooling energy.

The following sections describe the main alternatives.

3.1   Natural ventilation

Fresh air for building occupants is usually provided via a ducted system powered by a fan. The minimum amount of fresh air necessary to satisfy the occupants is supplied as a continuous flow into the building. Windows are sealed closed to prevent any unwanted ingress of air.

However, there may be periods when external air is cool enough to provide some useful cooling to the building. To make use of this potential low cost cooling source, a more flexible ventilation strategy is required.

Natural ventilation is the introduction of air into a building by natural means such as temperature differential or external wind pressure.

For example, by opening windows on either side of a building, external wind will encourage a cross flow of air through the building. Alternatively, high level vents will encourage air flow due to temperature rise. As entering air warms up, it will tend to rise to the top of the building from where it can be vented whilst new air is simultaneously drawn in at low level.

Natural ventilation can be used to provide air for ventilation purposes and to provide cooling. The circulation of cool external air on some summer days can be sufficient to cope with internal heat gains. Alternatively, natural ventilation can be used with night cooling to provide higher cooling outputs, as described in section 3.3.

A number of alternative control strategies can be considered to optimise the performance of a natural ventilation system including:

— *powered actuators*: to open and close windows and natural ventilation openings as required

— *wind speed sensors*: to limit or prevent window opening if wind speed is considered too high.

Natural ventilation

Summary

- Wind or temperature differences drive air flow through the building.
- Provides ventilation air for occupants and air for cooling.
- The building layout must be planned to maximise air flow rates and ensure good distribution.
- Automatic openable windows are used to control air flow patterns
- Can be used with night cooling.

Advantages

- Reduces the need for mechanical cooling (especially when used with night cooling).
- Avoids the need for a mechanical ventilation system (fans and ductwork).
- Avoids the energy consumption of fans and reduced cooling plant energy consumption.

Disadvantages

- Noise or contaminated air could enter the building.
- It may be more difficult to secure the building against break-ins.
- Nuisance draughts may cause disruption to occupants (disturbance of papers etc.)
- Careful control regime required to ensure proper performance and to maximise benefits.
- Difficult to ensure that internal spaces stay within fixed temperature limits.
- No control of internal humidity levels (only a problem if internal zones generate humidity).
- Performance can be difficult to predict — dynamic thermal and air movement simulations are required.

- The main thermal mass (structure) of the building is cooled using cold night air, enabling it to absorb more heat during daytime thereby reducing daytime cooling loads.
- Concrete floor slabs can be used for this purpose.
- Night air temperatures must fall below internal comfort temperatures.
- Ventilation air must make good contact with the slab and the slab must be in contact with room air to be effective.
- Purpose made decking systems can increase the reduction in cooling load achievable.

Advantages

- Reduces or avoids the need for full air cooling plant — the estimated offset in heat gains can be in the range 20–30 W/m² for night cooling with air.
- Can be retrofitted in existing non-air conditioned buildings.

Disadvantages

- All of those which apply to natural ventilation or mixed mode ventilation as described in the preceding two sections.
- Difficult to ensure that internal spaces stay within fixed temperature limits.
- No control of internal humidity levels (only a problem if internal zones generate humidity).
- Careful control of ventilation is required to avoid over-cooling.

— *wind direction sensors*: to optimise window opening around the building such that only windows on the windward and leeward sides of the building are opened.

In winter, uncontrolled air infiltration would allow heat to escape and waste energy. Therefore automatic opening windows must be able to seal shut during these periods. Openings must also be designed to prevent the ingress of rain and to maintain the security of the building.

3.2 Mixed mode ventilation

Mixed mode ventilation is a system which combines natural ventilation with a mechanical system of fans and ducts to achieve a more predictable solution. There are three common design approaches:

— Contingency designs incorporate both natural ventilation and mechanical ventilation so that each system can be used as a back-up to the other depending on the circumstance.

— Complementary designs function with both systems operating together, either concurrently or in changeover mode when one system closes down and the other starts up depending on conditions.

— Zoned designs allow for natural or mechanical systems to be used in different zoned parts of the building.

In comparison with natural ventilation systems, mixed mode ventilation has the following advantages:

— better able to maintain internal temperatures within required limits

— reduced risk of cold air drafts

— greater flexibility in building layout; layout does not have to be dictated by natural ventilation requirements

— less need for controlled window and other natural ventilation openings

— larger internal heat gains and cooling loads can be accommodated

— potential for fresh air heat recovery in winter using mechanical ventilation plant.

3.3   Night cooling with natural or mixed mode ventilation

Night cooling by fresh air ventilation is used to lower the temperature of the building structure at night when outside air temperatures are relatively low. By cooling the structure of the building at night the requirement for daytime mechanical cooling can be reduced.

During night time, the main exposed fabric of the building is cooled thereby removing heat absorbed during the day. During the following day, the structure is then able to absorb more heat than would otherwise have been the case, resulting in a delayed heat gain to the space and a consequently slower rise in internal temperature.

Night cooling is particularly effective in the UK where night time temperatures are nearly always below daytime comfort temperatures. The solution is well suited to buildings with mainly daytime cooling loads such as offices.

Figure 1 illustrates the main principle.

To be efficient, night time ventilation air must be in direct contact with the main exposed structure (thermal mass) of the building which, during daytime, must also be in direct contact with the internal air.

Concrete structural floor slabs and beams offer an ideal thermal mass which can be used as a thermal store. However, there must be good contact between the ventilation air and structural surfaces. Coverings with high

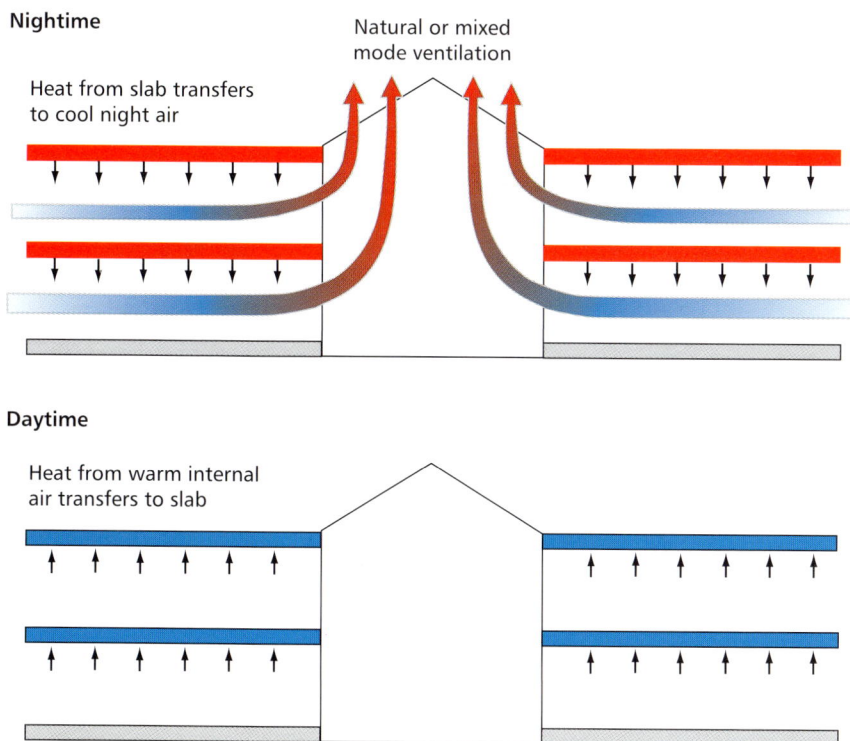

Nightime

Natural or mixed mode ventilation

Heat from slab transfers to cool night air

Daytime

Heat from warm internal air transfers to slab

Figure 1:
Night cooling strategy

thermal insulation properties such as carpets or false ceilings would substantially reduce the effectiveness of the thermal mass. Perforated ceilings are available which are more suitable for this purpose.

To improve the effectiveness of night cooling, internal slabs can be shaped to maximise surface contact with ventilation air. Additional cooling capacity can be achieved by using slabs with a honeycomb of internal ducts through which cool ventilation air is passed. After cooling the slab in this way during the night, the operation of the system can be changed such that during daytime, entering fresh air is pre-cooled by passing it through the same ducts before being supplied to the space or further conditioned. Such systems can achieve up to 50 W/m² of cooling. An example of this type of slab is shown in Figure 2.

Specially designed decking systems are now available which are able to achieve significantly greater levels of thermal storage than the slab on its own. Such systems can improve thermal transfer between the slab and ventilation air by encouraging fast turbulent air flow across the surface of the slab. Further capacity can be achieved by incorporating phase change materials (i.e. materials that will freeze at night and melt during the day) within the decking. These are sometimes able to store as much cooling as the slab itself.

Night cooling can be used in conjunction with either natural ventilation or mixed mode ventilation as described in the preceding sections. Mixed mode ventilation generally enables higher degrees of cooling, improved control and smaller air intakes but the benefits have to be compared against the additional fan energy required.

Figure 2:
Hollow slab to improve effectiveness of night cooling

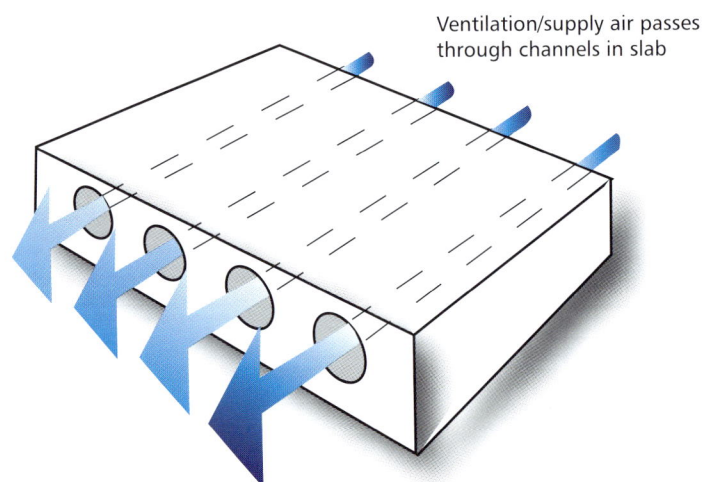

Ventilation/supply air passes through channels in slab

4 Cooling energy from renewable sources

The most readily accessible forms of direct cooling from renewable energy sources are from either the ground or local lakes and sea water.

In the UK, the annual swing in mean air temperature is around 20 °C whereas the mean temperature swing of the ground is around 8 °C or less depending on depth. This means that during summertime, below ground levels are lower than surrounding air temperatures whereas during winter they are often higher. This temperature difference can be used as a source of summer cooling or winter heating.

Similarly, groundwater, lake or sea water takes much longer to warm up than the surrounding air. Water at depth will remain particularly cold relative to the hot sunshine at its surface. This means that water extracted from depth can be used directly for cooling.

The following sections describe the main design options for cooling utilising these sources.

4.1 Ground cooling air systems

Ground cooling air systems work by drawing fresh air into the building via an underground matrix of ducts buried below the building. This process can be used to pre-cool ventilation air in summer, or to provide some degree of pre-heating during winter.

For example, air drawn in on a summer day can be cooled from 28 °C down to around 17 °C. In winter air can be heated from below zero temperatures to around 5 °C. These temperature changes represent a significant reduction in heating and cooling loads for the building. Ground temperatures vary depending on the depth and time of year. Usually systems buried to a depth of 2–4 m give the best economic return without incurring prohibitive excavation costs.

A typical ground cooling system is illustrated in Figure 3.

Since the surrounding ground is heated as warm air passes through the system, there will be a gradual reduction in the cooling output of the system if operated continuously. Ideally, intermittent operation should be planned, coinciding with occupancy periods. This will enable the system to be switched off for periods allowing heat transferred to the ground to dissipate.

Ground cooling

Summary

- Cooler temperatures below ground, or in water sourced from boreholes or deep lakes, is used for cooling.
- Systems can be for air, whereby ventilation air is pre-cooled by passing it through a network of underground ducts.
- Systems can be for water, whereby water is pre-cooled by circulating it through an underground coil. Winter heating can also be provided utilising heat pumps.
- Open loop systems use borehole water or lake water to cool chilled water directly.

Advantages

- Systems are usually simple to design and generally feature established, low maintenance components.
- There is little requirement for plant space inside the building.
- The cooling source is generally consistent and readily available.
- Used with heat pumps, cooling source is controllable and can also provide winter heating.

Disadvantages

- Excavation costs may be high depending on type of system, location and building situation.
- There may be restrictions placed on the use and extraction of ground water.
- Water must be taken from deep to be viable.
- Water taken from lakes and sea water may be corrosive.
- Filters and pumps require careful maintenance to avoid dirt related failures.

Figure 3:
Ground cooling air system

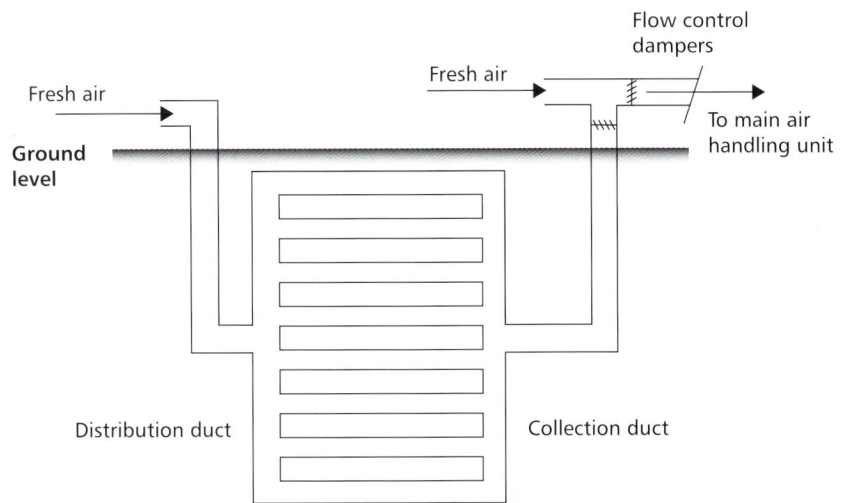

Fresh air

Fresh air

Flow control dampers

To main air handling unit

Ground level

Distribution duct

Collection duct

4.2   Ground cooling water systems (closed loop)

Closed loop type ground cooling water systems circulate water (or anti-freeze treated water) through a matrix of buried pipes. Typically a large continuous loop of flexible high density polyethylene pipe is buried in the ground adjacent to the building through which water is circulated.

There are two common configurations in which buried pipes are arranged:

—   *Vertical boreholes*: these are typically around 130 mm in diameter and extend up to 100 m deep. A single loop of pipe is inserted into each hole, then backfilled with high density grout. Vertical boreholes can provide very good performance due to their depth. This is further enhanced if there is a natural movement of groundwater across the pipe loop.

—   *Horizontal loops*: these are installed in trenches approximately 2 m deep. Although cheaper to install than vertical boreholes, the surrounding ground temperature will not be as low and hence, a larger area of buried pipe will be required. Coiled pipes are sometimes used for this purpose to maximise the buried pipe length for a given area of trench.

Due to the relatively small temperature difference between circulating water and the surrounding ground temperature, the cooling output of this type of system is limited. Therefore, reverse cycle heat pumps are commonly used in combination with closed loop systems as a means of increasing and controlling the amount of cooling or heating.

A heat pump is essentially a refrigeration machine which transfers heat from one environment to another. Used with a closed loop ground cooling system, during summer, the heat pump would extract heat from the building supply air or chilled water, and would transfer it to the surrounding ground via the closed loop pipework. Alternatively in winter, the same system can be operated to transfer low grade heat from the ground to the building typically via a low temperature heating system such as underfloor heating.

4.3   Ground cooling water systems (open loop)

Open loop type ground cooling water systems take water straight from the ground for cooling purposes. There is therefore no need for heat transfer through a buried pipe.

The system usually comprises two vertical boreholes, drilled to a suitable depth into an underlying aquifer (i.e. a layer of water-bearing rock which readily transmits water to wells and springs). Water is then extracted from one of the boreholes, passed through a heat exchanger, then re-injected back into the ground via the second borehole.

Typical groundwater temperatures are in the range 6–10 °C and, via the heat exchanger, are able to achieve chilled water temperatures of around 12 °C. Chilled water at this temperature is suitable for direct supply to chilled ceilings and chilled beam systems or as pre-cooling for a fan coil system. A typical open loop system is shown in Figure 4.

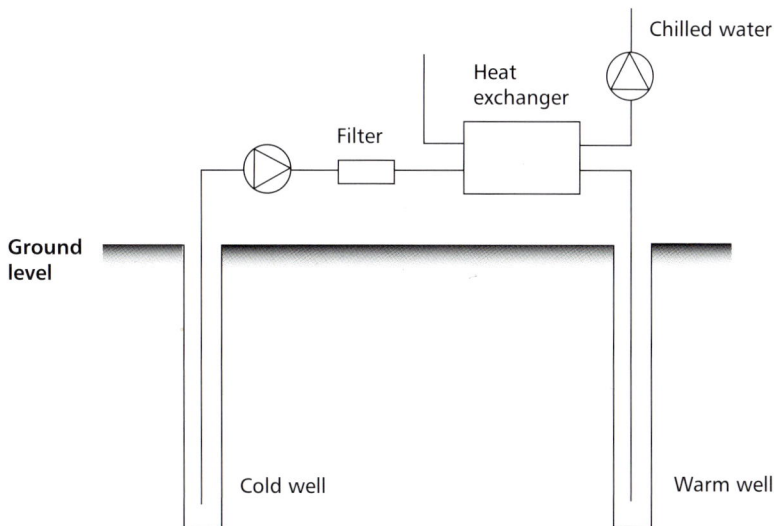

Figure 4:
Ground cooling water system

Although thermally very efficient, an open loop system runs the risk of blockage due to dirt and debris circulating in the water. Therefore some degree of filtering of the borehole water is required to minimise this risk.

Furthermore, an Environment Agency licence must be obtained for the abstraction and use of groundwater for this purpose.

4.4   Surface water cooling

Alternatives to the use of borehole water include lake water or sea water. For these systems water is extracted from as deep a location as possible, where the water is coldest. After passing through the heat exchanger the warmed water is passed back to the water source at high level, see Figure 5. For these systems, there needs to be a significant depth of water to ensure that extracted water is cold enough to provide useful cooling.

Figure 5:
Surface water cooling system

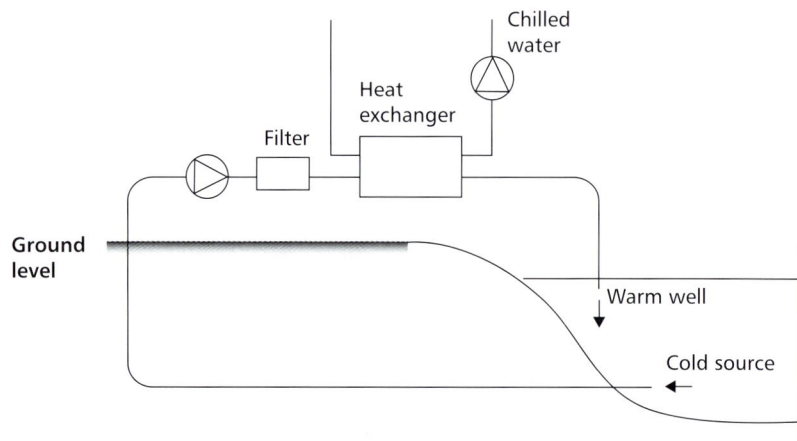

5   Free cooling options for use with mechanical cooling plant

Free cooling systems take advantage of favourable weather conditions to enable the chiller plant to be shut down for long periods. The incorporation of a free cooling capability to a mechanical air conditioning system can significantly reduce the annual energy requirement for cooling.
The technology involved is neither new nor complex, but opportunities to make use of it are often overlooked.

The viability of free cooling is increased if:

— there is a significant cooling load during winter months, typically greater than 20% of full design cooling load

— there is a continuous 24-hour demand for cooling

— chilled water can be circulated at increased temperatures

— a high performance building envelope is provided.

The following sections describe some of the most common solutions for achieving free cooling.

5.1   Free cooling in re-circulating air systems

In a system where air is conditioned at a central air handling unit and distributed via a constant volume ductwork system, it is likely that the total volume of air required to cool the space is greater than the amount needed to provide fresh air for occupants. This means that a proportion of the extracted air can be re-circulated back into the occupied space.

For example, in Figure 6 it can be seen that fresh air is mixed with extracted room air, conditioned, and then re-supplied back into the occupied space. The fresh air coming into the building and the waste air leaving the building are at the minimum amounts sufficient to provide fresh air ventilation to occupants — typically less than 25% of the overall amount of air required to cool the space.

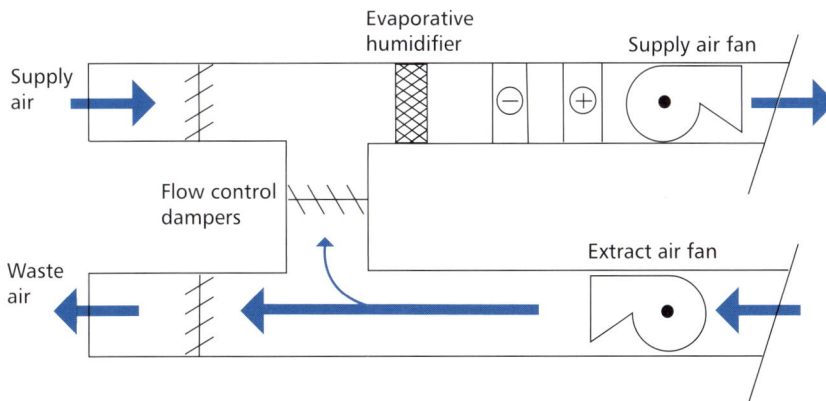

Figure 6:
Control of fresh air volume to achieve free cooling

To achieve free cooling, modulating dampers can be used to increase the ratio of fresh air to re-circulated air when the outside air temperature is less than the required internal space temperature. The supply, re-circulation and extract dampers can be controlled in unison to provide an increased proportion of fresh air as the first stage of cooling. As the outside air temperature increases, the dampers will modulate until the minimum fresh air volume is reached. Beyond this condition, cooling is achieved by operation of the mechanical cooling plant.

A similar result can be achieved for the heating system during winter months. Free cooling/heating of this type is now a standard feature of many packaged air handling units.

The use of a humidifier in systems with humidity control can provide further free cooling due to evaporation. By passing the mixture of fresh air and re-circulated air over an evaporative humidifier a useful degree of additional free cooling can be achieved. When water is sprayed or injected into the fluid stream, energy from the air is removed as the water droplets are evaporated. This can result in a temperature drop of up to 6–7 °C in the supply air temperature.

5.2   Free cooling in full fresh air systems

In systems where re-circulation of extracted air is not possible because the air is contaminated with pollutants such as smoke or fumes, it is now common practice to install some form of heat recovery device to transfer heat from extracted air to incoming air during winter, or from incoming air to extracted air during summer. Typical heat recovery devices include run-around coils, thermal wheels or plate heat exchangers.

The energy savings of such systems can be increased by the inclusion of free cooling in the form of an evaporative humidifier upstream of the extract air entry into the heat recovery device. Such a system is shown in Figure 7.

Due to evaporation of water droplets, the extracted air can be cooled by 6–7 °C before entering the heat recovery device. This means that free cooling is possible whenever the outside air temperature exceeds the extract air temperature after humidification. For some buildings, this can represent a significant proportion of the operating period of the system.

Specially manufactured air-to-air plate heat exchangers, which have water sprayed directly onto one side of the plates, are available to achieve free cooling of this type in full fresh air systems. Because the water is sprayed on one side of the heat exchanger, evaporative cooling is achieved without the addition of moisture to the air. However, such systems require careful treatment to prevent scaling or fouling of heat exchanger surfaces.

Figure 7:
Evaporative humidification to achieve free cooling

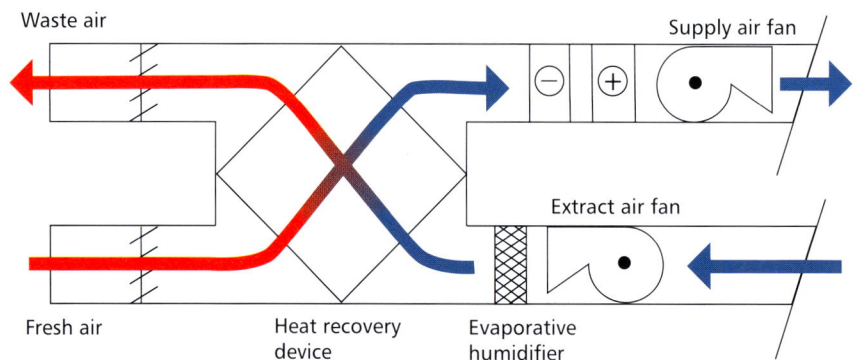

Waste air — Supply air fan — Fresh air — Heat recovery device — Evaporative humidifier — Extract air fan

5.3   Desiccant cooling systems

The free cooling capacity of a full fresh air system can be further improved by the addition of desiccant cooling. A desiccant is a substance that adsorbs moisture from the air. By using a desiccant to dry the air before passing it through an evaporative humidifier, a greater degree of cooling is achievable under a wider range of conditions.

Desiccant material is commonly supplied within a wheel which rotates through the incoming supply air. The desiccant wheel dries the air down to a relative humidity typically less than 20%. The air is then cooled by means of a heat recovery device, often a thermal wheel, which transfers cooling from the extracted air to the supply air. The dry, cooled, fresh air is then further cooled by passing it through an evaporative humidifier. Because of the ability of the air to evaporate moisture has been maximised, the degree of cooling is also maximised resulting in a temperature drop of 8–9 °C. Using this process it is possible to cool outside air at 30 °C down to around 17 °C ready for distribution into the building. A typical system is shown in Figure 8.

Although the process of cooling avoids the need for refrigeration plant, energy is still consumed due to the need to constantly dry and 're-generate' the desiccant material. Where possible, this heat could be provided by waste heat from some other process or solar energy. Nevertheless, even with electric powered regeneration, it is estimated that overall cooling costs can be up to 30% less than for mechanical chiller plant.

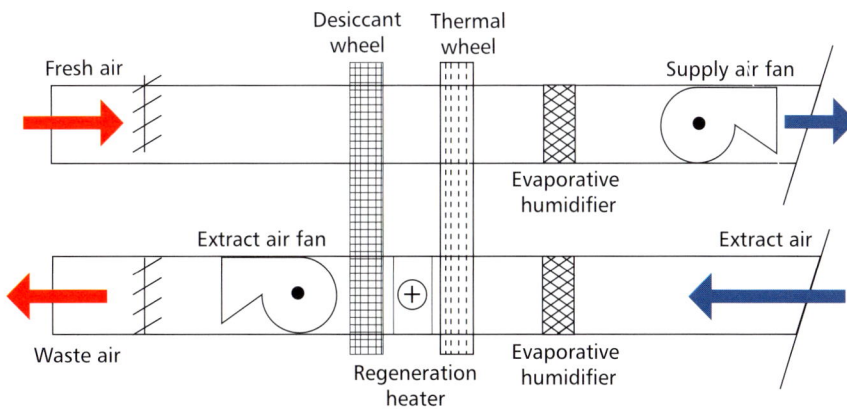

Figure 8:
Desiccant cooling solution

5.4   Free cooling in systems with cooling towers

A common cooling solution is to circulate chilled water to room terminal units such as fan coil units or chilled beams. For such systems, free cooling is achievable where heat rejection from the chillers is via cooling towers.

In such a system, chilled water would normally be cooled by circulating the water through the evaporator element of the chiller. The heat extracted would then be removed by the condenser element of the chiller which in turn would be cooled by the cooling tower. Evaporator and condenser elements are so called because refrigerant gas is being either evaporated or condensed as part of the refrigeration cycle.

There are therefore two separate water circulation systems; the chilled water system, circulating water between the evaporator and the terminal

Summary

- When outside air temperatures are low the system uses the cold air to cool the building thereby enabling the chiller plant to be shut down for long periods.
- Cool fresh air can be supplied direct to occupied spaces, or can be used to cool chilled water.
- Evaporative humidifiers can be used to enhance the degree of cooling achieved.
- Desiccant cooling can be used to further enhance the performance of evaporative humidifiers.
- Cooling towers and evaporative coolers can be used to provide direct cooling for chilled water.

Advantages

- Reduces the operating time for mechanical cooling plant thereby saving energy.
- Relatively cheap to implement.
- Maximum cooling requirements can still be met, i.e. mechanical plant can deal with periods of maximum demand to ensure internal temperatures are maintained.
- Some free cooling features can be purchased as part of a packaged air handling unit or chiller.

Disadvantages

- Design consideration needs to be given to determine the approach best suited to the building.
- Buildings need to have all year round cooling load to maximise periods of free cooling.
- Additional controls required to ensure proper performance and to maximise benefits.

units, and the condenser water system circulating water between the condenser and the cooling tower.

Free cooling essentially works by rejecting heat from the returning chilled water straight to atmosphere, by-passing the chiller completely. Free cooling can be achieved in this way during periods when the outside wet bulb temperature is less than the required chilled water temperature. For chilled beam systems in particular, where chilled water temperatures of around 14–15 °C are acceptable, free cooling is available for a significant proportion of the year. Some typical system layouts are shown in Figure 9.

The most efficient way to achieve free cooling is to by-pass the chiller such that chilled water is circulated straight to the cooling tower/evaporative

Direct system

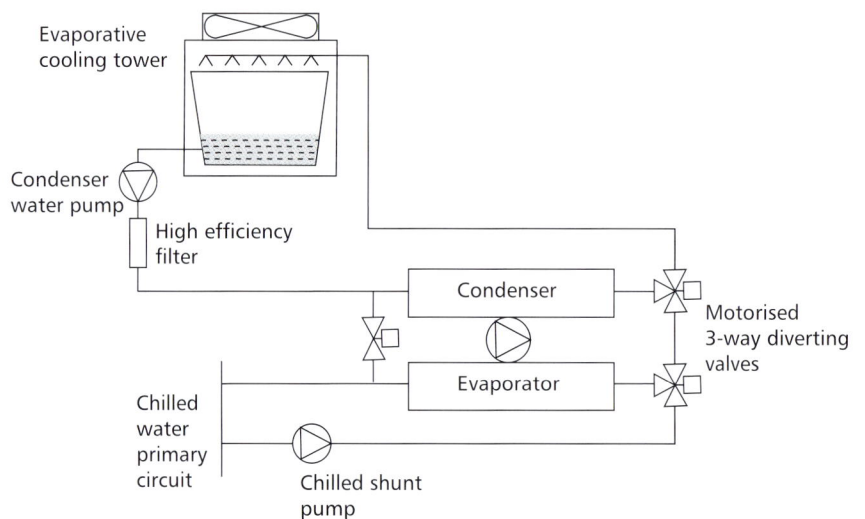

Indirect system

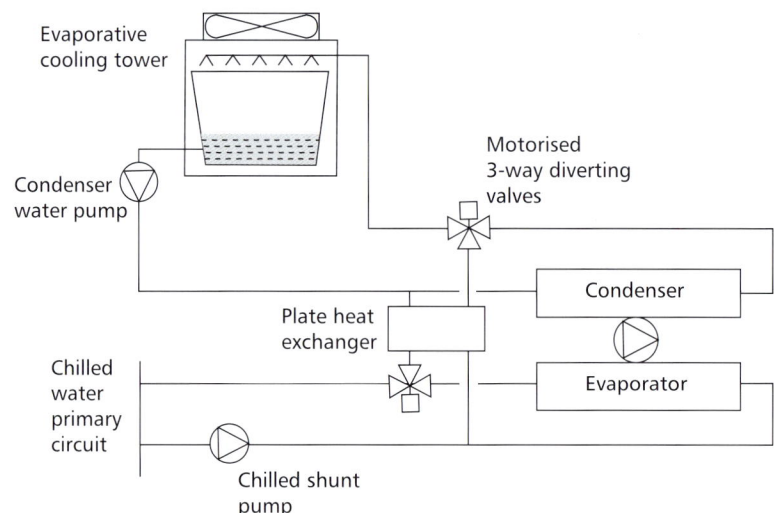

Figure 9:

Cooling tower free cooling solutions

cooler. This type of system is referred to as a 'direct system' and means that during free cooling, water from the chilled water system is allowed to mix with water from the condenser water system. This approach leaves the chilled water system exposed to the risk of contamination from debris entering the open condenser water system and therefore requires careful filtration and treatment of system water.

To avoid this risk chilled water and condenser water systems can be connected by means of a plate heat exchanger. This type of system is referred to as an 'indirect system'. Heat from the chilled water is rejected by circulating the water through a plate heat exchanger, the other side of which circulates water from the condenser system. Some manufacturers market combined open-circuit cooling towers with integral plate heat exchangers for this purpose.

The positioning of the three-port valves mean that, under free cooling, the chiller plant can be either by-passed completely (full free cooling) or used only to supplement the free cooling ('load shaving').

5.5   Free cooling in systems with evaporative coolers

Water-side systems with evaporative coolers are able to take advantage of the same free cooling method as for systems with cooling towers as described in the preceding section.

The difference between a cooling tower and an evaporative cooler is that for an evaporative cooler, condenser water is circulated through a finned coil across which air is drawn to carry away the heat. To improve heat transfer, the coils are kept wet by spray nozzles injecting water. Evaporative coolers have the advantage compared to cooling towers in that there is no need for a collection sump, and less water in contact with the air.

5.6   Simultaneous air and water free cooling

Chilled beam or fan coil systems where fresh air is supplied by a central air handling unit can take advantage of simultaneous air-side heat recovery and water-side free cooling.

During periods when the outside air temperature is less than the required chilled water temperature, the chiller is shut down so that returning chilled water is circulated through the cooling coil on the air handling unit. Under this condition, instead of providing cooling to the entering fresh air, the entering fresh air actually cools the chilled water and is slightly heated as a result.

For a fan coil system, outside air temperatures would need to be in the range 5–8 °C in order to provide cooling to returning chilled water temperatures of around 12 °C (assuming a chilled water flow temperature of 6 °C). However, if this technique is only used during winter months, it may be acceptable to allow a higher chilled water temperature in order to permit free cooling for longer periods. This would be possible if winter time cooling loads were significantly less than peak summertime loads, as would usually be the case.

For chilled beam systems where chilled water design temperatures are typically around 15 °C, free cooling in this way is possible at external temperatures of around 14 °C to suit chilled water return temperatures of around 20 °C.

An evaporative humidifier could be used to further increase the time when free cooling is available by lowering the dry bulb temperature of the entering air before it reached the cooling coil.

A typical solution is shown in Figure 10.

Figure 10:

Use of cold fresh air to pre-cool chilled water in winter

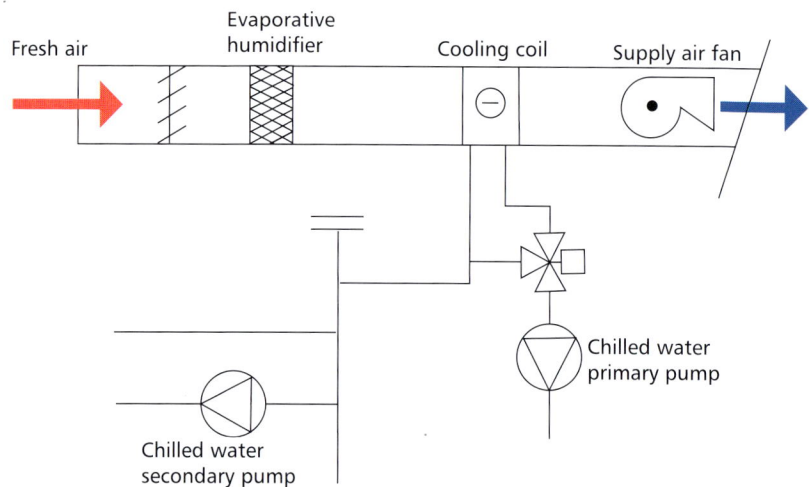

5.7   Thermosyphon systems

Thermosyphon systems can offer a packaged solution for water-side free cooling by incorporating free cooling within the chiller operation itself.

When external conditions permit, the difference in temperatures inside the condenser and evaporator are utilised to encourage natural circulation of the refrigerant around the circuit without the need to run the compressor. In this mode, refrigerant by-passes the compressor and enters the condenser where it is cooled. The resulting condensate then passes straight back to the evaporator, by-passing the expansion valve.

6　Delivery systems to maximise potential of low energy cooling

The means of delivering cooling to occupied spaces can have a significant impact on the effectiveness of the various low energy cooling options described in the preceding sections of this publication.

In general, delivery systems which can operate with higher than usual air or chilled water temperatures tend to offer much greater energy saving potential than those that require lower supply air or chilled water temperatures.

For example, terminal units that require chilled water at around 6–8 °C supply temperature are usually dependent on mechanical refrigeration plant with reduced opportunity for low energy cooling options. Terminal units or cooling surfaces which can achieve adequate cooling output with chilled water temperatures of 14–16 °C are far more likely to make low energy cooling solutions viable. Similarly, a cooling system which supplies air at 18–20 °C offers more low energy cooling potential than a system which supplies air at 15–18 °C.

The following sections describe those systems which are best able to maximise the benefits of low energy cooling.

6.1　Water cooled slabs

Cooling to the building is delivered by radiation from concrete floor slabs incorporating an embedded plastic pipe circulating chilled water.

Such systems can be used to supplement a night cooling ventilation strategy (see section 3.3) enabling higher cooling capacities to be achieved. Alternatively, the slab cooling can be operated during daytime occupancy periods to provide additional cooling during peak load conditions when a night cooling strategy alone may not cope.

A typical water cooled slab design is shown in Figure 11. The solution typically comprises a polyethylene or polybutylene pipe buried within the slab. Chilled water at around 14–20 °C is circulated through the slab during summer resulting in cooling capacities of up to 80 W/m²; chilled water any cooler than this could result in condensation on concrete surfaces.

Concrete has the ability to store cooling energy meaning that cooling input can be provided during night time and used to absorb heat gains during the daytime. This means that if a chiller is required, the installed chiller capacity is likely to be smaller than for a conventional building since it does not need to

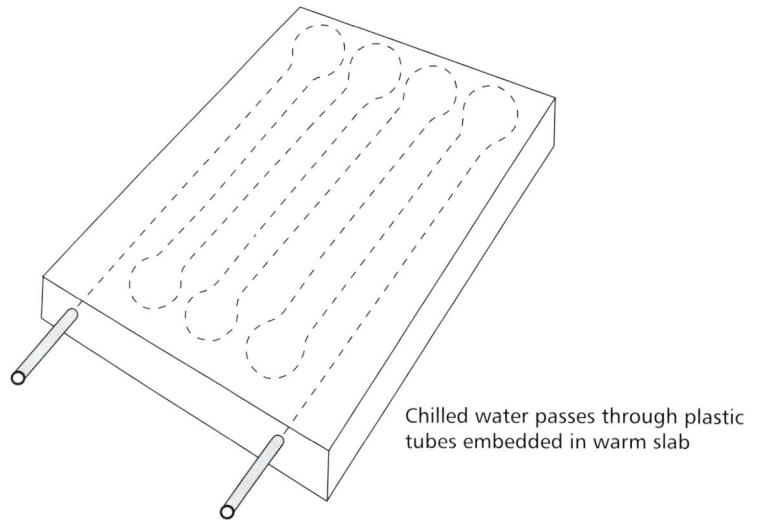

Chilled water passes through plastic tubes embedded in warm slab

be sized for the simultaneous peak cooling load. Furthermore, since the chiller would only be run at night, cheap rate electricity can be used.

6.2   Chilled beams and chilled ceilings

Chilled beams are available as terminal units that can be mounted above a false ceiling. They comprise long rectangular sections containing a finned tube through which chilled water at around 15–16 °C is circulated. Warm air rising up from the occupied space passes over the beams and is cooled causing it to drop back down into the space. Where necessary, fans can be incorporated to increase air movement across the tube.

Figure 12:

Chilled ceiling and chilled beams

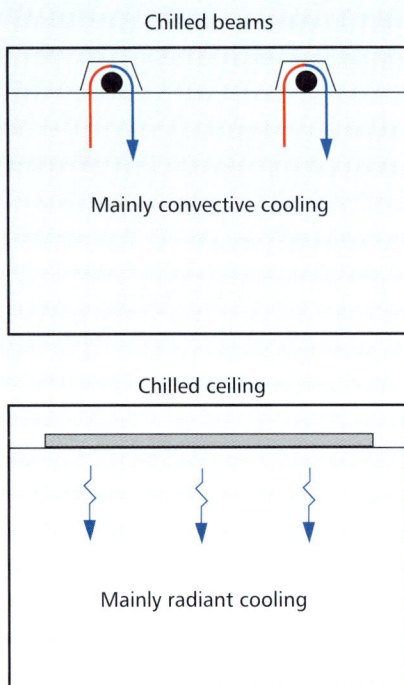

Chilled beams

Mainly convective cooling

Chilled ceiling

Mainly radiant cooling

Chilled ceilings comprise a small bore chilled water pipe arranged in a serpentine coil and fixed to a thin metallic panel. The panel is cooled through contact with the pipe and radiates cooling to the occupied space.
For both chilled beams and chilled ceilings, the cooled surface area is sufficient to achieve significant cooling outputs without the need to use low chilled water temperatures.

For example, fan coil units are smaller in size but require chilled water at around 6–8 °C supply temperature. This means that mechanical cooling is unavoidable. Chilled beams and chilled ceilings, see Figure 12, are able to provide equivalent cooling with chilled water temperatures of around 15–16 °C which means that low energy cooling options become viable.

6.3   Displacement ventilation

Displacement ventilation systems supply cooled air to the occupied space via low level wall mounted or floor mounted outlets. The cooled air tends to remain at low level until it comes into contact with a local heat source within the space such as a person or an item of equipment. It is then heated and rises to high level where it is extracted via a high level extract system.

Unlike more conventional ventilation systems this approach is deliberately designed to avoid the mixing of incoming supply air with room air. Instead, the supply air displaces the room air. This means that its cooling capacity is supplied directly to those parts of the occupied space where it is needed.

As a result, the supply air does not need to be cooled as much as would be the case in a conventional system. Conditioned air at approximately 18–20 °C (i.e. 2 K below the ambient room temperature) is usually sufficient. This compares with fresh air temperatures of 15–18 °C which are more common in conventional systems. This increased supply air temperature means that low energy cooling solutions and free cooling options are more viable. A displacement ventilation system can typically deliver 30–40 W/m² of cooling. This may not be adequate for a modern office building, and it is common for displacement ventilation to be supplemented by a high level cooling system such as chilled beams or a chilled ceiling.

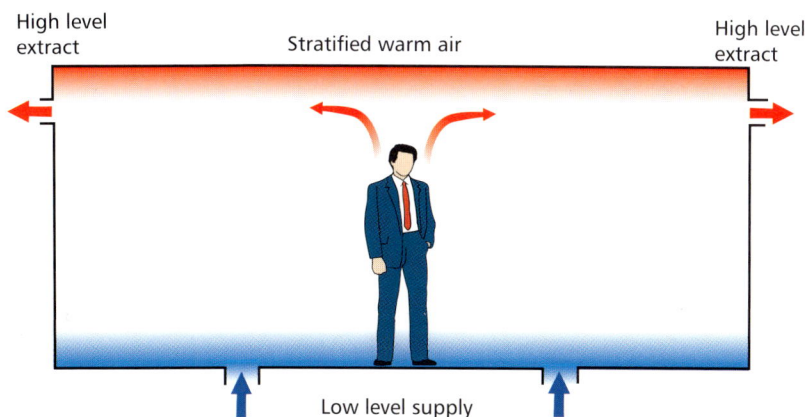

High level extract   Stratified warm air   High level extract

Low level supply

Figure 12:
Displacement ventilation

7 Selected bibliography

BSRIA *Free cooling systems* BSRIA BG8/2004 (Bracknell: Building services research and Information association) (2004)

Conservation of fuel and power Building Regulations 2000 Approved Document L (London: The Stationery Office) (2002)

Carbon Trust *New ways of cooling — Information for building designers* General Information Report GIR85 (Action Energy) (2002) (www.thecarbontrust.co.uk)

CIBSE *Climate change and the indoor environment: impacts and adaptation* CIBSE TM36 (London: Chartered Institution of Building Services Engineers) (2000)

CIBSE *Energy efficiency in buildings* CIBSE Guide F (London: Chartered Institution of Building Services Engineers) (2002)

CIBSE *HVAC strategies for well-insulated airtight buildings* CIBSE TM29 (London: Chartered Institution of Building Services Engineers) (2002)

CIBSE *Mixed mode ventilation* CIBSE AM13 (London: Chartered Institution of Building Services Engineers) (2002)

CIBSE *Ventilation and air conditioning* CIBSE Guide B2 (London: Chartered Institution of Building Services Engineers)

Liddament M W *Low Energy Cooling: Technical Synthesis Report* IEA ECBCS Annex 28 (Birmingham: FaberMaunsell/Energy Conservation in Buildings and Community Systems) (2000)